Human Factors in Flight
Student Workbook

Craig S. Funk

This *Student Workbook* is designed to help you identify and master the key concepts in the *Human Factors in Flight* textbook. The major part of the *Student Workbook* contains objectives and questions which focus on the key issues and concepts covered in the class. The textbook, written by Frank H. Hawkins, is excellent text material and you will note the *Student Workbook* follows the textbook very closely.

The **Performance Objectives** (POs) listed at the beginning of a section of questions are designed as student performance objectives. If you can accomplish the actions stated in the POs, you will have mastered the appropriate content.

The **Questions** questions are designed to assist you in focusing your study of the course material in preparation for participating in the regular class. They are also designed as discussion questions and most of them will be used as a springboard for the class discussion sessions. Appropriate space is provided between the questions to take notes on the answers discussed in the text and in the class. You can also use them to help you assess your mastery of the course objectives. If you can give the correct answer for each question, you are well on the way to accomplishing those objectives, and should do well on answering the examination questions designed to measure your achievement in the class.

The workbook also contains some **Supplemental Information** on Human Factors Investigation and Video Involvement Questions for selected videos used to illustrate various subjects in the course.

First published 1995 by Ashgate Publishing

Published 2017 by Routledge
2 Park Square, Milton Park, Abingdon, Oxon OX14 4RN
711 Third Avenue, New York, NY 10017, USA

Routledge is an imprint of the Taylor & Francis Group, an informa business

Most questions in this Workbook relate to the main textbook: *Human Factors in Flight*, Second Edition, 1993; © Frank Hawkins 1987, © Uniepers 1993; published by Avebury Aviation/Ashgate Publishing Company.

An *Instructor's Guide* which complements this Workbook is also available. Please contact the publisher for further details.

ISBN 13: 978-0-291-39831-4 (pbk)

Table of Contents

Background to Human Errors

PO: *1.*   *Describe the general impact of human factors on the aircraft accident record in recent years.*

1. Why is the study of Human Factors so important in the promotion of aerospace safety?

2. How does the human record look in the **worldwide commercial jet fleet** accident record?
   (From Boeing's 1993 Statistical Summary)

	1959 - 1993	LAST 10 YEARS
Flightcrew as a Primary Factor All Phases	_____%	_____%
Flightcrew as a Primary Factor Final Approach & Landing	_____%	

3. What impact does the human have on the accident record of the **U.S. Air Carriers**?
   (From 1991 NTSB Annual Review p 18)

Broad Cause/Factor	All accidents 1991	1986-1990	Fatal Accidents 1991	1986-1990
Pilot	____%	____%	____%	____%
Other Person (not aboard)	____%	____%	____%	____%
Other Person (aboard)	____%	____%	____%	____%

4. What extent is the human involved in the accident record of **U.S. general aviation** aircraft?
   (From 1992 NTSB Annual Review)

Broad Cause/Factor	(p 22) All accidents 1992	1987-1991	(p 34) Fatal Accidents 1992	1987-1991
Pilot	____%	____%	____%	____%
Other Person (not aboard)	____%	____%	____%	____%
Other Person (aboard)	____%	____%	____%	____%

5. How does "Personnel" fit in as a broad cause or a factor in the 1992 **general aviation** aircraft accidents?
   (From 1988 NTSB Annual Review p 39)

   - _____%   = Personnel

   - _____%   = Pilot

6.   What do all these statistics boil down to in terms of how many accidents have the human as a contributing cause?

Worldwide Commercial Jet Fleet
All Phases of Flight
Flightcrew as a Factor   _____
1983 - 1993

U.S. Air Carrier
Pilot as a Broad Cause Factor   _____
1983 - 1987

Worldwide Commercial Jet Fleet
Final Approach & Landing Phase
Flightcrew as primary factor   _____
1959 - 1993

General Aviation "Personnel"
as a Broad Cause Factor   _____
in 1992

PO: 2.   *Give a brief historical perspective on the development of human factors.*

7.   When and with who did human factors as a technology start?

8.   What were the important milestones in the first century of human factors development?

9.   When did serious interest get started in generating a greater awareness of human factors amongst those responsible for design, certification and operation of aircraft?

10.   What did the International Air Transport Association (IATA) conclude about human factors in their 20th Technical conference in 1975 at Istanbul?

11.   What are the two broad principles that must be accepted to achieve better application of human factors in civil aviation?

PO: 3.   *Define the meaning of human factors.*

12.   What is human factors about?

13.   What might be some key words in a good definition of human factors?

(Taken from *Human Factors for General Aviation*, p 1-2)

14.   What is the applied technology of human factors concerned with?

15.   What are the twin objectives of human factors applications?

16.   What does the term "ergonomics" mean?

17.   How is human factors different from ergonomics?

18.   What were some of the early misconceptions about human factors?

19.   What problems, in addition to the physiological, must human factors be concerned with?

20.   How can you describe human factors as a "multi-disciplinary" technology?

The SHEL Conceptual Model

PO: 1.   *Describe the SHEL conceptual model of human factors in terms of what each of the letters mean.*

1.   What do each of the letters in the SHEL conceptual model of human factors represent?

S =   _____

H =   _____

E =   _____

L =   _____

2.   What is the most valuable and flexible component in the system?

PO: 2.   *Describe with examples the various characteristics of the **LIVEWARE** component of the system.*

PO: 3.   *Identify the scientific discipline which is associated with each characteristic of the liveware component.*

3.   What are the various characteristics of the liveware component to consider in the human factors conceptual model?

PO: 4.   *Describe each of the interfaces of the SHEL human factors conceptual model and give some examples of considerations which apply to each of them.*

4.   What must be taken into account in order to make more effective **LIVEWARE-HARDWARE** (L-H) interface in the SHEL conceptual model?

5.   How does the **LIVEWARE-SOFTWARE** (L-S) interface in the SHEL conceptual model work?

6.   Which interface in the SHEL conceptual model was one of the earliest recognized in the flying environment?

7.   What are some of the examples where the L-E interface is trying to adapt the human to fit the environment?

8.   What are some of the ways the L-E interface is trying to adapt the environment to fit the human?

9.   What other L-E consideration is generated in human factors with the speed of transmeridian travel and payload flying 24 hours per day?

10.   What are the considerations of the **LIVEWARE-LIVEWARE** (L-L) interface of the SHEL conceptual model of human factors?

11.   What does human factors attempt to research and explain?

The Nature of Error

PO: *1.*   *Be able to describe the nature of human errors in terms of normal distribution and other factors which may affect the distribution of accidents.*

1.   What insight did the Roman orator Cicero give into the nature of human error?

2.   What kind of human error may result from a person reacting in a perfectly normally manner to the situation presented?

3.   What kind of characteristics do errors like these have?

4.   What things did the Three-Mile Island nuclear incident, the Chernobyl nuclear disaster, and the Tenerife double 747 crash have in common?

5.   What are the three basic tenets with respect to human error?

6.   What does the author consider to be one of the most enduring myths in aviation?

7.   How are pilot errors different from any other human error?

8.   What fallacies are implied by use of the term "pilot error"?

9.   If human error is normal, of what value is the human operator?

10. If human error is considered to be a part of norman human behavior, what question must be answered to plan accurately for man in the system?

11. What are some normal human error rates?

PO: 2. *Describe the term "accident proneness" and its usefulness in mishap prevention efforts.*

12. How would you define the term "accident proneness"?

13. How can random distribution account for some people having more accidents than others?

14. What might be reasons, other than normal distribution, which could account for some people having more accidents than normal?

15. How do we reduce inherent awkwardness and lack of muscle coordination as reasons for greater than normal accident distribution?

16. What seems to have the greatest effect on the short term factor of simple carelessness?

17. What are the short term influences which may cause a person to have more potential for mishap, and which them presents the most promise for a fruitful approach to prevention?

18. How does the short term influence of stress affect the human?

Sources of Error

PO: 1. *Explain and give examples of how mismatches between the SHEL components can be a source of human errors.*

1. Using the SHEL model, where are the potential areas of mismatch between the components?

2. What are some examples of the kind of problems resulting from a poor interface between Liveware-Hardware?

3.   What areas of mismatch might come up in the Liveware-Software interface?

4.   What are some of the conditions which lead to increased errors in the Liveware-Environment interface?

5.   How can deficiencies in the interface between the Liveware-Liveware components develop?

PO: 2.   Describe the errors which can occur in the information processing system of the Liveware component in terms of sensing, perception, decision-making, action and feedback.

6.   What are the areas that are the sources of error within the information processing system of the liveware component?

7.   How do each of these fit together in the human information processing system?

8.   What might be the causes of the individual differences in the sensing process?

9.   What vital-to-flight information is the human sensory system not designed to detect?

10.   What is perception?

11.   What are some of the Gestalt laws which are concerned with structure or organizational arrangement of elements?

12.   Which organizational arrangement is the basis for all perception?

13.   How does the context of the message affect the way it is perceived?

14.   What happens to perception when there is ambiguous or inadequate information?

15.   What influence does "set" or expectation have on perception?

16.   How does the practice of standardizing equipment and procedures fit into the laws of perception?

17.   How can the law of expectation have a detrimental effect on errors?

18.   Under what conditions is reversion to an earlier habit pattern more prevalent?

19.   What complicates the relationship of reliability (freedom from errors) to experience?

PO: *3.*   *Describe how motivation affects performance and what conditions might reduce the motivation level.*

20.   What is motivation and what effect does it have on the performance of a particular task?

21.   What factors might bring a person's motivation down?

PO: *4.*   *Explain how arousal or alertness relates to how well humans perform.*

22.   What is the relationship between the degree of arousal and alertness to the effectiveness of a person in performing a task?

23.   What happens to performance when arousal is higher or lower than optimum?

24.   How does the optimum level of arousal for speed differ from that required for accuracy?

PO: *5.*   *Describe the factors which affect the human decision-making process.*

25.   What are the factors which can distort the human decision-making process?

8

PO: 6. *Explain how the false hypothesis or mistaken assumption contributes to error in the human decision-making process.*

26. What is the most dangerous characteristic of the false hypothesis?

27. What are the situations where the mistaken assumption is more likely to occur?

28. What is the primary drawback to the human decision-making process?

29. How can the action process in the human component be adversely affected?

30. How important is feedback to the human information processing system?

PO: 7. *Describe some of the conditions which lead to errors in eye witness accounts of accidents.*

31. What are some of the possible influences which can affect the accuracy of eye witness accounts of accidents?

32. What things can impact on the perceptions of the eye witness?

33. What communication factors can be a source of error in the testimony of the eye witness?

34. How does the issue of retention fit into the accuracy of a witnesses' testimony?

35. How might the nature of the event influence the testimony of the eye witness?

36. What precautions should be taken to increase the accuracy of eye witness testimony?

Error Classification and Reduction

PO: *1.*   *Describe the four ways to classify errors.*

1.   What are the four different ways to classify human errors?

2.   Which interfaces of the SHEL model are usually involved with design-induced errors?

3.   What is the cause of an operator-induced error?

4.   What are the differences between the random, systematic and sporadic errors?

5.   How would you define an error of omission?

6.   What is an example of an error of commission?

7.   How would you describe an error of substitution?

8.   What is useful about classifying errors as reversible and irreversible?

9.   What is an example of an error that is reversible?

PO: *2.*   *Describe some of the differences between human and machine in tasking to reduce errors.*

10.   What is the general principle used to allocate the tasks between human and machine?

11. What are the tasks that are best matched to human and machine performance?

    Machine: Human:

12. What did the Royal Air Force (RAF) laboratory tests in 1943 conclude about the human's performance of tasks requiring the monitoring or detection of brief, low intensity and infrequent events over long periods of time?

13. When does the "vigilance decrement" phenomenon usually occur in human performance?

14. Since practice does not seem to be effective, what is the best way to eliminate the phenomenon of vigilance decrement?

15. How can the deficiency in human short term memory be reduced?

PO: 3.   *Explain and give examples of the elements in the two-pronged attack on reducing human errors.*

16. What are the two prongs of the two-pronged attack on human errors?

17. Which of the two prongs seems to have the most relevance in terms of system design?

18. What basic premise must be accepted to implement the second prong of the attack on human error?

19. What does minimizing the occurrence of error focus on?

20. What does minimizing the occurrence of errors mean in the context of the SHEL model?

21. What are some examples of the SHEL interfaces to consider in minimizing the occurrence of error?

    a. L-H

    b. L-S

    c. L-E

    d. L-L

11

22. In the SHEL context, what matching is required to minimize the occurrence of errors?

23. How does designing for minimizing the occurrence of errors fit in with the level of arousal?

24. How does the classification of errors as random, systematic or sporadic fit into the attack on minimizing the occurrence of errors?

   a. Random

   b. Systematic

   c. Sporadic

25. What are some of the principles connected with the second prong of reducing the consequences of errors?

Fatigue, Body Rhythms, and Sleep

PO: 1.   *Describe the causes of jet lag and fatigue and how these phenomena affect human performance.*

1. How would you define the term "Jet Lag"?

2. What are the more specific performance symptoms of jet lag?

3. What are the dangers for flying while under the influence of jet lag?

4. How prevalent in the air carrier industry is it for flight crews to be fighting jet lag?

5. How does the Galvanic Skin Response (GSR) system assist in countering some of the effects of jet lag?

6.   How big is the problem of fatigue in the aviation industry?

7.   What indications are there for greater formal education in the human factors associated with fatigue?

8.   What are the four causes of fatigue?

9.   What are some other conditions that can contribute to fatigue?

PO: 2.   Describe some of the body rhythms and how they relate to human factors of performance.

10.   Which of the body rhythms is the most significant?

11.   When did the first chronobiological reports in the literature appear?

12.   What serves as the focal point of scientific study in chronobiology?

13.   What are some of the cycles of body chemistry which coincide with the 24 hour oscillations?

14.   What does the term acrophase have reference to in chronobiology?

15.   Which single circadian pacemaker or controlling mechanism governs the body's biorhythms?

16.   What are some of the rhythm and time cues that maintain the 24 hour cycle, and what are they are called?

17.   How does the oral temperature rhythm indicate the daily cycle?

18.   What does the Birthdate Biorhythm Theory have to offer?

19.   What can be said about the rhythm of performance in the 24 hour cycle?

20.   What kind of an influence does the disturbance of biological rhythms have in aviation?

21.   What are some of the terms associated with the disturbance of biological rhythms?

22.   What are the four factors which are concerned with adapting to living environment changes?

23.   How has the development of chronohygiene drugs progressed?

24.   What things can be used to enhance the chronohygiene adjustment of the phase shift of the biorhythm?

PO: 3.   *Describe the characteristics of the two basic kinds of sleep and how sleep effects the physiological well being of the human.*

25.   What is the most common physiological symptom of long-range flying?

26.   What causes this disruption in the normal sleep pattern?

27.   How would you explain the difference between "monophasic" and "polyphasic" in describing sleep patterns?

28. What is meant by the "Golden age of sleep"?

29. What are the two basic kinds of sleep?

30. How does the pattern of sleep change with age?

  - At birth   = ___% Orthodox and ___% REM

  - Adulthood   = ___% REM

  - By age 70   = ___% REM

31. What are the four subdivisions of orthodox sleep?

32. What are the three factors affecting the recuperative effect of naps?

33. How can you apply the benefits of a nap?

34. What is microsleep, when does it usually occur, and what benefit does it bring?

35. What can be said about the quality of sleep?

36. What does the term "anchor sleep" refer to?

37. How does sleep affect memory?

38.   What are the two types of insomnia and the causes of each?

39.   What are the types and possible causes for clinical insomnia?

40.   What are the causes for situational insomnia?

PO: 5.   *Describe how some of the commonly used drugs affect sleep.*

41.   How do various drugs affect sleep?

Barbiturates
      &
Benzodiazepines (Valium):

Alcohol:

Caffeine:

Amphetamines:

42.   What is the function of sleep?

PO: 6.   *Describe the performance characteristics of a sleep-deprived person.*

43.   What are the performance characteristics of the sleep-deprived person?

44.   What operational considerations must be made with regard to sleep loss and human performance?

45.   How effective are drugs and hormones in helping with control of biorhythms?

46.   How is Autogenic Training set up to work?

47.   What benefit does Autogenic Training have on improving sleep?

48.   How can exercise affect sleep?

Fitness and Performance

PO: 1.   *Describe total and partial incapacitation, their causes, differences and the operational considerations for countering their effects.*

1.   What are the extreme pathological conditions which can lead to sudden total incapacitation?

2.   What is the cause of the more frequent in-flight incapacitation of pilots?

3.   What type of operational measures are currently being taken to minimize the risk of sudden incapacitation?

4.   From an operational standpoint, what is the difference between total and partial incapacitation?

5.   What are some of the reasons for partial incapacitation?

PO: 2.   *Define fitness and explain how physical fitness relates to mental fitness and overall human performance.*

6.   How would you define fitness?

7.   How important is a pilot's physical condition as it effects performance as a pilot?

8.   What is the connection between physical fitness and mental performance?

      - Physical fitness ---- > Mental health

      - Physical fitness ----- > Physical health

PO: 3.   *Describe how each of the six main factors which affect fitness, also effect human performance in the flying environment.*

9.   What are the main factors which affect fitness and performance?

<u>POSITIVE</u>   <u>NEGATIVE</u>

10.   What are the three general benefits attributed to fitness achieved by a suitable exercise program?

11.   How does a person who feels better from an exercise program benefit?

12.   What does looking better from a good exercise program do to benefit personal performance?

13.   In what ways is a person with a good exercise program actually healthier?

14.   What are the three types of exercises which should be included in the effective physical conditioning program?

15.   What is the best way to achieve this three fold kind of fitness?

16. How do some of the usual "exercise" activities provide for fitness?

17. What are the three significant components of tobacco smoke which destroy fitness?

18. What are the main health hazards statistically linked to smoking?

19. What effect does nicotine have on the body?

20. What effect does tar have on the body?

21. How does carbon monoxide affect health and fitness of the body?

22. What is the estimated cost of alcoholism in industry and the overall cost of alcohol abuse in the U.S. annually?

23. How does alcohol affect performance?

24. What are the insidious flying performance implications for alcohol users which may not be generally known?

25. What is the accident rate for those who abuse alcohol compared to other workers?

26. How does the blood alcohol content (BAC) relate to flying performance of a pilot?

27. What part does alcohol play in the aviation accident record?

28. What are some of the indicators a friend or supervisor might observe in a pilot who is tending toward alcoholism?

29. What part do on the job use of drugs by pilots play in aviation safety?

        Frequent or occasional   = ___%

        Rarely ....   = ___%

        Total ....   = ___%

30.   How prevalent are drugs in our society?

31.   What impact do drugs and alcohol have on the aviation accident record in recent years?

32.   What is a "stressor"?

33.   What are some of the adverse results of stress in the human?

34.   What are the **physiological** stressors the typical pilot has to face?

35.   What are some of the **psychological** stressors a pilot has to deal with?

36.   How does the Life Change Units (LCUs) system of measuring potential stress work?

37.   What are the two most commonly applied means of managing stress?

38.   Where does the damage from the stressor come?

39.   What are some of the more effective means to manage the response made by individuals to the stresses in their lives?

40.   How does the "Comprehensive Program" at work assist individuals in managing or reducing the stress in their lives?

41.   What is the keystone to the comprehensive program of stress management?

42.   What is the primary function of food in the diet?

43.   What are the seven types of intake the human body needs to function effectively?

44.   What are the three main types of food to provide the body with a balanced diet?

Vision

PO: 1.   *Define the various terms associated with the measurement of light.*

1.   What is the measurement of light called?

2.   What are the three basic terms used to determine if the lighting is sufficient?

3.   What does the term **contrast** refer to?

4.   What is meant by **refraction**?

5.   What causes the eye to see **color**?

6.   What is the function of each of the following parts of the eye?

   - **Pupil**

   - **Iris**

   - **Ciliary muscle**

   - **Lens**

   - **Cornea**

   - **Retina**

   - **Optic nerve**

   - **Blind spot**

   - **Rods**

   - **Cones**

   - **Fovea**

   - **Extrinsic muscle**

7.   How are the following terms used in describing the operations or functions of the human eye?

   - **Accommodation**

   - **Mandelbaum effect**

   - **Binocular vergence**

   - **Stereopsis**

   - **Photopic vision**

   - **Scotopic vision**

   - **Adaptation**

   - **Rhodopsin**

   - **Visual acuity**

PO: 3.   *Describe the elements which make up visual perception.*

8.   What are the main elements of visual perception?

9.   What happens when the perception process is denied prediction cues?

10.   Where does the processing of visual information get started?

11.   What other factors may also play a role in perception?

12.   Where in the perceptual process do uncertainty and ambiguity occur with correctly sensed information?

13.   What is **fascination** as it relates to the visual task?

14.   How does **set** play a role in visual perception?

15.   How does the **Critical Fusion Frequency** pertain to visual perception?

PO: 4.   Describe what causes the daytime "Blind Spot" and how normal vision compensates for it.

16.   What is meant by the "Blind Spot" in the eye?

PO: 5.   Describe how the perception of depth and distance are made using each of the various visual cues.

17.   What are the visual cues the eye and brain uses to establish perception of depth and distance?

PO: 6.   Describe the detrimental effects of smoking on vision.

18.   What are the detrimental affects that smoking has on vision?

Visual Illusions

PO: 1.   Be able to identify the ways visual illusions can come into visual perception and give examples of optical and depth/distance illusions.

1.   Why is it important for pilots and human factors specialists to know about visual illusions?

2.   What are the two areas in which visually sensed information can have errors induced?

3.   What are some of the examples of **optical illusions** common to visual perception?

4.   What are some examples of **depth and distance illusions** common to visual perception?

PO: 2.   *Define the following illusions:*

             - Brightness contrast   *- Autokinesis*
             - Somatogyral illusion   *- Oculogyral*
             - Somatogravic illusion *- Oculogravic*
             - Coriolis illusion   *- Induced movement*   *- False horizon*

5.   Where is the predominant area of incidents and accidents arising from visual illusions?

6.   What causes the **brightness contrast** illusion?

7.   How would you describe the cause of **autokinesis**?

8.   What is the **somatogyral** illusion?

9.   What causes the **somatogravic** illusion?

10.   What are the visual components of the somatogyral and somatogravic illusion called?

11.   How does the cross coupled or **Coriolis** illusion occur?

12.   What causes the illusion of **induced movement**?

13.   What can cause the perception of **false horizon**?

PO: 3.   *Describe the various illusions which can occur during pilot operations of aircraft.*

14.   What are some examples of illusions that can occur during taxying?

15.   What are the illusions characteristic to the takeoff phase of operations?

16.   To what illusions might a pilot be susceptible in cruising flight?

17.   What are some of the conditions which contribute to the illusions pilots can experience during the approach and landing phases of operations?

18.   What are some of the approach and landing illusions which can affect pilots?

a. Sloping terrain to the runway:

Condition		Illusion	Effect
- Up slope	=	_____	_____
- Down slope	=	_____	_____

24

b. Sloping runway:

Condition		Illusion	Effect
- Up slope	=	_____	_____
- Down slope	=	_____	_____

c. Runway width different than normal:

Condition		Illusion	Effect
- Wider	=	_____	_____
- Narrower	=	_____	_____

d. Light intensity:

Condition		Illusion
- Brighter	=	_____
- Dimmer	=	_____

e. Visibility restriction (Mist or fog):

Condition		Illusion
- Brighter	=	_____
- Dimmer	=	_____

f. Windshield Location - pitch angle:

Condition	Illusion
- high speed	_____
- low speed	_____

g. Runway texture:

- Lack of texture -	_____

PO: 4. *Describe five things which can be done to minimize flightcrew susceptibility to visual illusions.*

19. What are five things flightcrews can do to reduce their chance of experiencing visual illusions?

PO: 5. *Define the "design eye" in terms of cockpit geometry.*

20. Where does the standard for the required visibility envelope for the flight deck of civil aircraft come?

21. Where should the "design eye" be located for civil transport aircraft?

22. What sitting height position will place the pilot's eye at approximately the "design eye" location?

PO: 6. *Describe some problems which result from poor cockpit window design.*

23. How can the optical characteristics of windows affect visibility?

Motivation and Safety

PO: 1. *Identify when human behavior began to be an important part of American accident investigation.*

1. What is the reason that a properly qualified, highly trained, medically fit, well paid person would fail to perform the task expected?

2. When did the National Transportation Safety Board (NTSB) first establish a separate Human Performance Division?

3. What is necessary in order to accomplish the best matching of the Liveware component to all the other components of the SHEL model?

PO: 2. *Define what is meant by the term "motivation" and describe the different levels of motivation.*

4. What is probably the most significant characteristic of the Liveware component in driving a person to behave in a particular way?

5. How would you define motivation?

6. Where does the basic level of motivation begin?

7. What is the end result of a sequence of motivated behavior called?

8. What determines the way we go about satisfying needs?

9. What may be the result of setting an unrealistically high goal?

10. How important is organizational culture to the motivation required for crews to maintain high standards of professionalism and exercise proper discipline?

11. What are some aspects of an individual's motivational system?

PO: *3.*   *Describe the three basic theories of motivation, and show how Maslow's theory on the hierarchy of needs works in relation to human motivation.*

12.   What are the three basic theories of motivation?

13.   How does Maslow's theory on the hierarchy of needs work in relation to human motivation?

PO: *4.*   *Describe the contributions of the industrial studies of Taylor, Hawthorne, and the Two-Factor Theory of Herzberg to understanding the motivation behind behavior.*

14.   What do we learn about motivation from the industrial studies of Taylor, Hawthorne, and the Two-Factor Theory of Herzberg?

   a. Taylor -

   b. Hawthorne studies -

   c. Two-Factory Theory of Herzberg -

PO: *5.*   *Describe how Murray's Motives fit into the theories of motivation.*

15.   What are the premises which form the foundation of Murray's Motives?

16.   How would you describe the three extensively investigated motives cited by Murray?

   a. **Achievement motivation:**

   b. **Affiliation motivation:**

   c. **Power motivation:**

17.   What happens if it is impossible to satisfy these individual motivational needs?

PO: *6.*   *Describe some of the considerations of the concept of expectancy and rewards in motivation.*

18.   In what way did Vroom criticize the Two-Factor Theory of Herzberg?

19.   How does expectancy of rewards fit into motivation?

20.   What is the most important aspect of the availability of rewards?

PO: *7.*   *Describe factors which influence job satisfaction, and differentiate between job enlargement and job enrichment.*

21.   What are some of the factors which influence the **job satisfaction** part of motivation?

22.   How does job satisfaction tie into performance?

23.   What is the difference between the job enlargement and job enrichment approaches to increasing job satisfaction?

        a. Job enrichment:

        b. Job enlargement:

24.   What guidance has been given for having a well established relationship between goal achieving and job satisfaction?

25.   What is one of the most pervasive problems facing advanced industrial societies?

26.   What are the six categories, suggested by surveys of job attitudes, into which boredom may fall?

PO: *8.*   *Describe the difference between positive and negative methods of reinforcing or discouraging behavior.*

27.   What is the difference between positive and negative behavior reinforcement?

28.   What are some of the precautions that must be applied when one is required to use negative reinforcement?

PO: *9.*   *Describe the elements of crew resource management (CRM).*

29.   How important is it for good leadership to be applied in the management of cockpit resources?

30.   What is involved in CRM?

PO: *10.*   *Define the role of leadership, and describe the characteristics and tasks of a leader.*

31.   What is a leader?

32.   What are the tasks of the effective leader?

33.   What is the difference between leadership and authority?

34.   In what ways do all members of a group contribute to the effective leadership of the group?

35.   What are the characteristics of a leader?

Communication

PO: *1.*   *Give the definition and some examples of types of communication.*

1.   How important is effective communication to our society and to the aviation industry?

2.   What is the definition of communication?

3.   Where does communication fit in the SHEL conceptual model?

4.   What are some of the types of communications which must be optimized for efficiency and safety?

2.   *Describe the elements required for communication and the place language has in the process.*

5.   What is required to accomplish communication?

6.   Which of the elements of the communication process uses language?

7.   What is ambiguity as it relates to the language of communication?

8.   What are the principles or rules called which govern the arrangement of words in language?

9.   What are the differences in the roles of written and spoken language?

   a. Speech

   b. Writing

10.   How does body language fit into the communication picture?

PO: 3.   *Define the term intelligibility in spoken language and give examples of ways to increase it.*

11.   What is meant by the term intelligibility in spoken language?

12.   What are some of the factors that make a word more intelligible?

13.   What has been done to use these principles to improve communications in the aviation industry?

14.   How does the standard vocabulary effect the intelligibility of a message?

PO: 4.   *Describe the main parts of the vocal and auditory systems and their functions.*

15.   What are the main parts of the vocal system?

16.   What are the main parts of the auditory system?

PO: *5.*   *Describe the types of conditions which can lead to loss of hearing ability.*

17.   What are the three ways hearing loss can occur?

18.   What is interference with the transmission of sound waves through the outer and middle ear called?

19.   What are some of the things which can cause hearing deficiencies from conduction deafness?

20.   What are the conditions which can cause nerve deafness?

21.   What causes a central hearing loss?

22.   What happens to hearing as a result of age?

PO: *6.*   *Describe the four characteristics of speech which influence intelligibility.*

23.   What are the characteristics of speech which can have an effect on the intelligibility of the message?

PO: *7.*   *Describe how clipping, masking, and noise affect the spoken message.*

24.   What are the kinds of degradation which can effect the reception of the spoken message?

25.   How does hearing protection reduce the affects of noise?

26.   How do visual cues combined with auditory information affect message reception?

PO: *8.*   *Explain how expectation can influence the reception of the spoken message and what can be done to reduce this phenomenon.*

27.   What increases the risk of expectation causing an error in understanding the spoken message?

28. Where is the phenomenon of expectation particularly common and dangerous, and what can be done about it?

PO: 9. *Describe some of the means of measuring the effectiveness of spoken language.*

29. What is the use of an articulation index good for?

30. What is meant by the Speech Interference Level?

PO: 10. *Describe some of the considerations in the development and use of Automatic Speech Recognition.*

31. What is Automatic Speech Recognition (ASR), and what are some of the problems to be overcome in its development?

32. What are the three main issues which must be resolved in order to have effective ASR systems?

33. What is the essential design criterion for ASR to be used on the flightdeck?

PO: 11. *Explain some of the safety issues associated with radio communications, and describe some changes for its improvement.*

34. Why is it important for all pilots to use and understand a common language?

35. What impact do errors in understanding oral communication play in aviation safety?

36. What was the most critical error centered around in the Tenerife double 747 accident?

37. What changes in radio phraseology were made after the Tenerife double 747 accident?

Attitudes & Persuasion

PO: 1. *Define and describe the differences between personality traits, attitudes, beliefs, and opinions.*

1. How would you define a personality trait?

2.   How would you define an attitude?

3.   What is a belief?

4.   How would you describe an opinion?

5.   How can personality, attitudes and beliefs be studied?

6.   In the jargon of psychology, what would the terms personality, attitudes and beliefs be called?

PO: *2.*   *Define the developmental influences for, and the components of attitude.*

7.   Where do attitudes have their origins?

8.   What are the three components of attitude?

9.   What might be some of the difficulties in evaluating attitudes by behavior?

PO: *3.*   *Describe how stereotyping can effect the development of attitudes.*

10.   What is stereotyping in terms of attitudes?

11.   Why do people tend to use stereotypes?

12.   What is the danger of stereotyping?

PO: *4.*   *Describe some of the reasons for having attitudes.*

13.   What are some of the more common reasons for having attitudes?

PO: *5.*   *Describe and differentiate between the Thurstone, Likert, and Guttman scales for attitude measurement.*

14.   How does the researcher try to measure attitudes?

15.   What component of attitude is most directly measurable?

33

16. What is the best known way of trying to determine attitudes?

17. What are the characteristics of the Thurstone, Likert, and Guttman scales for measuring attitudes?

   a. Thurstone Scale:

   b. Likert Scale:

   c. Guttman Scale:

18. What is one way these attitude surveys or opinion polls may be distorted?

PO: 6. *Decribe, in terms of social psychology, the difference between a collection of individual people and a group.*

19. In what way is a group more than simply a collection of people?

PO: 7. *Describe some of the attitudes and behavior of individuals who belong to, and are being influenced by a group.*

20. What are three types of group influenced behavior which may have a detrimental effect on the individual performance with regard to safety?

21. What influence does a group decision have on the element of risk-taking?

22. What are the reasons a group has a tendency to accept greater risk than an individual?

23. What causes the loss of inhibition an individual experiences when in a group?

24. What are the factors which constitute the influence of a group on the individual to maintain conformity?

25. When are group pressures to conform less effective on an individual?

PO: *8.*   *Describe how attitudes might be affected by efforts to change them.*

26.   What makes a person's attitude resistant to change?

27.   Which attitudes are more easily changed?

28.   How can a person's resistance to changing their attitude be altered?

a. Decreased

b. Increase

29.   What can be done to improve undesirable personality traits in cockpit crew members?

30.   How effective will training be in changing the attitudes of a pilot?

PO: *9.*   *Describe the characteristics of communications designed to change attitudes.*

31.   What are the functions of communications?

32.   What is the most significant function of communication?

33.   What are the three basic parts of the persuasive communication process?

34.   What characteristics must the person have to be an effective user of persuasive communications?

35.   What can be done to enhance the persuasive effectiveness of the message itself?

36.   How does the nature of the receiver affect the effectiveness of persuasive communication?

37.   What are some of the reasons that any single persuasive safety communications does not seem as effective as we might hope?

Training and Training Devices

PO: *1.*   *Define the terms education, training, and skill as used in the text and differentiate between them.*

1.   What are some of the subjects included in the study of the human factors in training devices?

2.   How important is training to our lives in general and to aviation in particular?

3.   How does training fit into the SHEL conceptual model?

4.   How would you define **Education**?

5.   What is **Training**?

6.   Where does the term **Instruction** fit into the training concepts?

7.   What are **Skills**?

8.   How would you differentiate between knowledge and skill?

PO: *2.*   *Describe some of the alternatives to training.*

9.   What are some of the alternatives to investing in training programs?

PO: *3.*   *Define and explain the principles of training transfer, feedback, guidance, cueing, prompting, and pacing.*

10.   How would you explain the concept of **training transfer**?

11.   How would you define **feedback**?

12. How do you differentiate between the open and closed-loop feedback systems?

   a. **Closed-loop** =

   b. **Open-loop** =

13. What is the difference between intrinsic and extrinsic feedback?

   a. **Intrinsic** Feedback

   b. **Extrinsic** Feedback

14. How would you define and describe the difference between guidance, cueing and prompting?

   a. **Guidance** =

   b. **Cueing** =

   c. **Prompting** =

15. What does **pacing** refer to in the training scenario?

PO: 4. Explain the stages in the learning process, how memory can be improved, and some of the handicaps to learning.

16. What is the difference between learning and training?

17. What are the three phases of learning?

18. What are the two kinds of memory and the characteristics of each?

19. What are some of the common memory enhancement techniques?

20. What effect does sleep have on memory in the learning process?

21. What effect does age have on memory?

22. What are the things which can become a handicap to learning?

PO: *5.*   *Describe each of the various training systems, and explain the advantages and disadvantages of each.*

23.   What should be considered when selecting any given training system?

24.   What are the advantages and disadvantages of the **lecture**?

     a. Advantages:

     b. Disadvantages:

25.   What are the pluses and minuses of the **lesson**?

     a. Advantages:

     b. Disadvantages:

26.   What are the good and bad points of **discussions**?

     a. Advantages:

     b. Disadvantages:

27.   How does the **tutorial** system work for efficiency?

     a. Advantages:

     b. Disadvantages:

28.   What are the advantages and disadvantages of **audio-visual** methods?

     a. Advantages:

     b. Disadvantages:

29.   What are the values and drawbacks of **programmed instruction**?

     a. Advantages:

     b. Disadvantages:

30.   How would you describe the pros and cons of a system using **computer-based training**?

     a. Advantages:

     b. Disadvantages:

PO: 6.   *Describe what is entailed in the systems approach to training.*

31.   What is meant by using the systems approach in training?

PO: 7.   *Explain the difference between training aids and training equipment, and what considerations must be made for their procurement.*

32.   What are the two major classes of training devices ?

33.   What things must be considered when procuring training devices?

PO: 8.   *Describe the concept of fidelity as it relates to training devices and training efficiency.*

34.   What are the two powerful incentives for the development of simulators?

35.   What does the degree of fidelity refer to in describing a simulator's performance?

36.   How much fidelity is required for effective training?

37.   What SHEL interfaces should have high or low fidelity?

38.   What training situations may require high fidelity?

39.   When would a low fidelity device be preferable?

40.   What does psychological fidelity depend upon?

Documentation

PO: *1.*   *Describe how some aspects of language can effect the efficiency of written documentation.*

1.   What are some reasons it is important to have proper and correct documentation in the aviation industry?

2.   What are some of the requirements any technical writer must meet to produce an effective document?

3.   What part of documentation should display consistency?

4.   What are the three basic aspects which require human factors optimization?

5.   What is basic about the choice of language for written documentation?

6.   What are the primary problems which must be faced in good technical writing?

7.   Who is better qualified than the linguist or literary scholar to use the language and words of machines?

8.   How should writing skills be focused for communicating effectively with passengers?

9.   What are some of the general rules directed toward optimizing comprehension?

10.   What is "jargon," and what guidance is given for its use in technical documentation?

11.   How can you evaluate the reading level of technical material?

PO: *2.*   *Describe how various aspects of printing can effect the efficiency of written documentation.*

12.   What, in general, is being considered when evaluating printing for documentation efficiency?

13. How is print size considered in documentation?

14. What can be done to increase legibility of smaller print?

15. What are the two general distinctions used for type-face styles?

16. Which group of type-faces is best for documentation?

17. How does upper or lower case affect readability?

18. What effect do *italics* have on readability?

PO: *4.*   *Describe how various aspects of layout can effect the efficiency of written documentation.*

19. What is going to be the major factor for determining the layout of a document?

20. What type of documentation would use single vs double column layout?

21. How does justification effect the readability of text?

22. How should spacing for paragraphs and other elements be used?

23. What is an essential requirement before the mass publication of a technical document?

PO: *5.*   *Describe the purposes for using checklists in the cockpit of transport aircraft, and some of the ways they are used.*

24. What are some of the purposes for a checklist in the cockpit of a transport aircraft?

25. How are checklists used in the cockpits of transport aircraft?

PO: *6.   Describe some of the human factors principles used in effective design of technical documents.*

26. How important are page numbering, indexing and a table of contents for technical documents?

27. What is the recent, significant development for helping to present technical information?

28. What are the functions of illustrations in technical documentation?

29. What has contributed to the 50-80 percent human error rate causing system failures?

PO: *7.   Describe the principles used to enhance the effectiveness of questionnaires and forms as written documentation.*

30. What can questionnaires or surveys be used for?

31. What is the value of the survey dependent upon?

32. How should the survey program be designed to be most effective?

33. What is the most important objective in the design of the survey questions?

34. What are some of the types of questions that can be used in a questionnaire?

35.   How does the "order effect" influence the information derived from a survey?

36.   How can wording affect the outcome of the information received on a survey?

37.   What are the three "Ls" of form design?

PO: 8.   *Explain how the application of human factors can be used to improve the documentation found on maps and aeronautical charts.*

38.   What is the difference between a map and a chart?

        a. Maps

        b. Charts

39.   How has a systematic and professional application of human factors been applied to the field of chart documentation?

40.   What are the two basic reasons for neglecting an adequate consideration of human factors in chart making?

41.   What things need to be done to improve the human factors applications to charts and maps?

42.   What was used in the 1940s by ICAO as a basis for the standardization of the symbols which appears on aeronautical charts?

Displays

PO: 1.   *Describe the historical development of human factors in cockpit displays and controls.*

1.   What are the main sources of pressure for the developments in displays and controls in modern transport aircraft?

2.   What were some of the earliest flight instrument displays?

3.   What were some of the major historical milestones in the development of cockpit displays?

4.   What are the three necessary pillars upon which the progress of cockpit displays has been built?

5.   What essential pillar for proper development of displays has been frequently neglected?

6.   What were some important early human factors studies that lead the way to better design of cockpit displays?

7.   What individuals formed the vital bridge between operations in air transports and the aircraft manufacturing industry?

PO: 2.   Explain the SHEL interface for displays and controls and the associated human factors problems.

8.   How do the cockpit displays differ from the cockpit controls in the SHEL model?

9.   To what human senses are displays designed to transfer information?

10.   What is the purpose of a display?

11.   Where does the first major human factors problem arise in the mission of display technology?

PO: 3.   Describe and give examples of some of classifications and the general design aspects of visual displays.

12.   What are the various classifications of visual displays?

13. What are some general considerations for the design of cockpit displays?

14. How are visual different from aural displays with regard to their directional use in the cockpit?

15. How would you differentiate between the digital vs the analog visual display systems?

PO: *4.*   *Describe some of the factors associated with the visual display of alphanumerics.*

16. Which way should the mechanical drum rotate to display numbers?

17. What is happening to the mechanical and electro-mechanical alphanumeric displays recently?

18. What human factors concerns must be addressed in electronic displays?

PO: *5.*   *Explain the different types of dial markings and presentations, and the considerations for their use.*

19. What are the three basic types of dynamic displays?

20. What are some of the basic human factors principles to consider in designing display scales?

21. What are the factors which determine the size of the display?

PO: *6.*   *Describe some of the applications, advantages, and problems presented for cockpit displays in the use of cathode ray tubes (CRTs).*

22. Why is the use of CRTs in cockpit displays considered such a milestone?

23. What are the general applications of CRTs as cockpit displays?

24. How are CRTs applied in the Airbus and the Boeing 757/767?

25.   What are some of the challenges to consider with the use of CRTs as flight deck displays?

26.   What are some of the advantages and disadvantages of using the developing "flat panel" displays?

PO: 7.   *Describe some of the advantages and difficulties with the development and use of the head-up display (HUD).*

27.   What area of commercial aviation is the most promising for the use of the head-up display (HUD)?

28.   What are some of the applications of the HUD other than low visibility approaches?

29.   How does a HUD work?

30.   Which part of HUD technology has had the most development work?

PO: 8.   *Differentiate between the fail-passive and fail-operational design concepts used in automatic flight control systems.*

31.   What is the difference between a fail-passive and a fail-operational automatic landing system?

       a. Fail-passive:

       b. Fail-operational:

32.   What are the primary reasons for developing and installing the HUD in aircraft flying in bad weather?

       a. ALPA:

       b. Airlines:

PO: 9.   *Describe the Warning, Alerting, and Advisory systems and explain the differences between them.*

33.   What seems to be the trend in the development of warning systems?

34.   To what kinds of things are the warning/advisory systems designed to draw the pilot's attention?

35.   What is the general concept established by SAE in 1980 with regard to warning systems?

36. What are the three fundamental objectives for application in the design of all flight deck warning systems?

37. What is paramount in the design of warning systems?

38. What are the four functional classes of alerting systems?

39. How are the alerting signals grouped for priority?

40. What are the advantages and disadvantages of the synthesized voice?

   a. Advantages:

   b. Disadvantages:

41. What is one of the essentials for cockpit displays with regard to the reliability of a display system in event of system failure?

Controls

PO: *1.*   *Describe the various functions of the controls in the cockpit and give examples of each.*

1. How do aircraft controls fit into the SHEL model for human factors interface?

2. What are the functions a control may fill?

3. How do controls differ in the forces required to actuate them?

PO: *2.*   *Explain each of the five design principles for cockpit controls.*

4. What are the five principles of design for cockpit controls?

PO: 3. *Explain the applications of keyboards in the cockpit.*

5. What are the developments in cockpit technology which bring the use of keyboards as controls to the forefront?

PO: 4. *Describe the differences between the QWERTY and DSK or Dvorak keyboard layout.*

6. How would you describe the differences between the QWERTY and DSK or Dvorak keyboard layout?

a. QWERTY -

b. DSK or Dvorak

PO: 5. *Explain the design considerations for flight deck applications of keyboards.*

7. What are the flight deck applications of keyboards which must be considered in the design?

8. What may be the role of the keyboard in the cockpit of the future?

PO: 6. *Describe the advantages and disadvantages of using automation in the cockpit.*

10. What are the three broad objectives (advantages) cited to justify the introduction of cockpit automation?

11. What has been one of the more notable human factors problems arising from the development of automatic systems?

12. What are some of the advanced concepts for the modern cockpit controls of today's, and future air transport aircraft?

Space and Design

PO: *1.*   *Describe the aspects of human factors to apply in the design of space in transport aircraft.*

1.   What is the primary tasks of the human factors specialist in designing working and living space in transport aircraft?

2.   What aspects must be considered in this optimization?

3.   What areas need to be considered for space and layout design?

4.   What knowledge must be applied when designing equipment for use on aircraft?

PO: *2.*   *Define the terms anthropometry and biomechanics and cite applications.*

5.   What are the two disciplines involved in matching hardware, software and the environment to the characteristics of the human?

6.   What is anthropometry concerned with?

7.   What is the study of biomechanics concerned with in human factors?

PO: *3.*   *Explain the various aspects of human dimensions as they relate to aircraft design considerations.*

8.   How is information collected about human dimensions?

9.   What care must be taken by the researcher when establishing a standard for human dimensions in the design of equipment?

10.   What else must be considered when designing for human dimensions?

11. What other differences must the designer of equipment for worldwide use consider?

12. What is the U.S. transport category aircraft design requirement for accommodating crew size ?

PO: *4.*   *Describe how the study of statistics works to aid the human factors considered in aircraft space design.*

13. What statistical concepts have value in arranging the results of anthropometric studies into meaningful design guidance?

14. What are the two main indices required for the description of the design population?

15. What are the general human factors guidelines used for space design?

16. Where does the statistical concept of percentile fit into the design for space in aircraft?

PO: *5.*   *Describe some of the constraints imposed on the proper human factors design of cockpit and cabin space.*

17. What has been the difficulty in the past with the design of the cockpit space?

18. What constraints often limit the extent the designer can optimize hardware?

19. How do visibility requirements influence design for space?

20. What considerations must be made for designing the space between the pilots?

    a. Closer

    b. Farther

PO: *6.*   *Describe the historical development and principles of design for layout of cockpit display panels.*

21.   What are the types of cockpit display panels that need to be considered for human factors optimization?

22.   What are the cockpit geometry concerns for panel displays that are associated with viewing distance?

23.   What were the historical developments in improving the design of cockpit instrument panels?

24.   What design considerations should be given to the design of the system quantitative information?

25.   What systems often use the mounting of displays and controls in a schematic or synoptic form?

26.   What are some of the basics for design of the flight guidance control panels?

27.   What other control and display panels in the cockpit or elsewhere require optimization?

28.   What are the two methods for determining the direction for toggle switch movement?

29.   How does the sight and reach principle apply to cockpit display and control panels?

PO: *7.*   *Describe the way the crew complement on an aircraft influences the layout of the flight deck.*

30.   How does the crew complement influence the layout and design of the flight deck?

PO: *8.*   *Describe design considerations for the crew seats on transport aircraft.*

31.   What are the requirements which drive design considerations for flight deck crew seats?

32.   What considerations should be given to cabin crew seat and facility design?

Human Factors in the Cabin

PO: 1.   *Describe the human factors to consider in the design of the cabin environment in terms of considerations given for the passengers and the crew.*

1.   How important are human factors in the cabin?

2.   What areas of the cabin must be considered for human factors optimization?

3.   What fundamental difference is there concerning the cabin environment, between the cabin crew and the passengers?

4.   What governs the priority of considerations between cabin crew and passengers?

5.   How does anthropometry and biomechanics fit into cabin human factors design?

6.   What should be the primary motivation for the flight attendant?

7.   What is one of the great challenges of working with people in the cabin?

PO: 2.   *Describe the major regions of the cabin requiring human factors optimization.*

8.   What is a major cabin region demanding human factors attention in matching the hardware to the liveware?

9.   Upon what is efficiency in the galley dependent?

10.   What does delethalization refer to?

11.   What special human factors considerations must be made when designing emergency equipment?

12.   What does the inadvertent actuation of an emergency slide cost?

13.   What are the problems associated with designing emergency equipment?

PO: *3.*   *Describe factors associated with the use of medical kits on board transport aircraft.*

14.   What are the two priority actions a flight attendant should take in the event of a medical emergency in the cabin?

15.   What are the most troublesome conditions occurring to passengers in flight?

16.   How frequent are in-flight deaths?

PO: *4.*   *Describe the causes of the fire hazards associated with the cabins of transport aircraft and corrective actions being taken to reduce them.*

17.   What are the major hazards in crash survivable accidents?

18.   What requirements are now in force to reduce the toxic gasses in aircraft fires?

19.   What actions must be taken to reduce the loss of life due to cabin fire?

PO: *5.*   *Explain the important human factors considerations for cabin seat design.*

20.   Why is the cabin seat so important to a passenger airline?

21.   What seat design considerations must be made?

22.   What is seat pitch?

23.   What G loading requirements do seats have?

24.   What increase in G load force is derived from upper body restraint?

25.   What requirements do cabin attendant jump seats have?

26.   What is the human factors difference between forward and rear facing seats?

27.   What other furnishings in the cabin must also conform to human factors design considerations?

PO: *6.*   *Give examples of software items in the cabin which require human factors optimization.*

28.   What are some examples of the items of cabin software requiring human factors optimization?

PO: *7.*   *Describe the broad categories of cabin duties and explain the human factors issues to be considered in them.*

29.   Into what two broad categories are cabin staff duties divided?

30.   Who in the cabin seems to suffer more injuries from in-flight turbulence?

31.   Why do cabin crew suffer more of the in-flight turbulence injuries?

32.   How important is the neatness and general appearance of the cabin staff in the Liveware-Liveware interface?

PO: *8.*   *Identify each of the cabin environmental areas and describe how human factors is applied to each.*

33.   What are the factors of concern in improving the interface between the Liveware and Environment components of the cabin?

34.   What are the sources of noise in fixed wing transport aircraft?

35.   What are the noise sources in helicopters?

36.   What are the effects of noise on those subject to the environment of transport aircraft?

37. What human factors considerations arise from temperature and humidity aspects of the cockpit/cabin environment?

38. What are the human factors of pressurization to consider in the environment of the cabin?

39. What are the human factors considerations for ozone in the cabin environment?

40. What human factors issues on aircraft come from smoking?

41. What considerations in the cabin must allow for the circadian and time zone effects?

Interface Between People

PO: 1.   *Identify some of the considerations that must be given to communications in the cockpit and cabin.*

1. How important is communication in the activities of the cabin?

2. What are some of the barriers to communications between the flight deck and the cabin?

3. What type of problems can decrease the effectiveness of the public address (PA) system?

4. What unique position does any crew member seem to be in when they walk down the aisle?

5. What special physical handicaps to communication in the cabin must be accounted for?

PO: 2.   *Describe the requirements for dealing with intoxicated passengers.*

6. What are the regulatory requirements with regard to intoxicated passengers?

PO: 3.   *Describe the considerations associated with a passenger's fear of flying, and cite measures which can help reduce this human response.*

7.   How does fear of flying or flying phobia fit into the cabin Liveware-Liveware interface?

8.   How can you best treat the fear of flying?

9.   What benefits can come from recognizing and accounting for fear of flying?

PO: 4.   *Describe the aspects of passenger behavior problems in the cabin, and give practical measures to deal with them.*

10.   What passenger behavior in the cabin can have serious safety implications?

11.   What are some practical measures that can be taken against passenger abuse/assault of the flight crew?

PO: 5.   *Describe the nature of passenger behavior in an emergency, and what crew members can do to enhance successful emergency response.*

12.   What are the characteristics of flying that are unique with regard to behavior in an emergency?

13.   What important Liveware-Liveware principles are highlighted from the feelings flying passengers have about their situation?

PO: 6.   *Identify the factors highlighted by the statistics related to survivability.*

14.   What survivability statistics come from analysis of aircraft accidents?

15.   What areas of concern remain with regard to important factors in survivability?

PO: 7.   *Describe the three classes of people who might be encountered, and what the primary tasks of the crew might be in a violent hijacking situation.*

16.   What are the three classes of people who might be encountered in a violent hijacking?

17.   What is the primary task of the crew during a hijacking?

Education and Applications

PO: *1.*   *Describe why there seems to be a barrier between the knowledge obtained in human factors research and the practical application of human factors principles.*

1.   What are the reasons for the communication breakdown between the academia of human factors and its application?

PO: *2.*   *Describe the basic educational preparation for human factors work and the specific qualifications for each of the four levels of needed human factors expertise.*

2.   What background seems to be the most common for human factors specialists?

3.   What are the four levels of qualifications for working in human factors?

Level 4

^^

Level 3

^^

Level 2

^^^

Level 1

^^^
4.   What are the criteria used in the development of the KLM Royal Dutch Airlines human factors awareness course?

PO: *3.*   *Describe some of the specific human factors training which airlines can develop and what they might be designed to accomplish.*

5.   Why might an airline develop a specific human factors training program?

6.   What are some of the recent team training and testing concepts in the air transport industry?

7.   How would you describe Line Oriented Flight Training (LOFT)?

PO: *4.*   *Explain the concept of Crew Resource Management (CRM), describe what it involves, cite what it is designed to accomplish, and specify what it will not do.*

8.   What is involved in CRM?

9.   What are the two general assumptions which are at the origin of CRM?

10.   What are the elements of CRM?

11.   What can CRM do?

12.   What will CRM not do?

PO: *5.*   *Describe how applications of human factors should be made in the six stages of design and production.*

13.   In what areas must human factors be applied to receive optimum effectiveness in the air transport industry?

14.   Who is responsible for making sure human factors is applied in the design and production of resources within the air transport industry?

15. What are the six stages of aircraft hardware design where human factors must be applied?

1-   4-

2-   5-

3-   6-

PO: *7.*   *Describe how human factors expertise is designed into training, operating procedures, staff/management relations, accident investigation, the regulatory authority, and marketing.*

16. How does the design of training fit into the human factors picture?

17. Who has responsibility for designing human factors into operating procedures and documentation?

18. How can human factors be designed into Staff/management relations?

19. How does human factors fit into accident investigation?

20. What part does human factors expertise have in the regulatory authority?

21. How important is an understanding of human factors to marketing?

PO: *8.*   *Describe some of the difficulties in "selling" the ideas of using human factors, and list some of the benefits which can come from a future with better applications.*

22. Why must the presentation techniques and material used to present human factors concepts employ the standards of human factors?

23. Where do the origins of resistance to human factors progress often lie?

24. How are the benefits of human factors applications best quantified?

25. What are some of the benefits of better application of human factors in aircraft operations?

26.   What is the key to success in human factors?

Human Factors Investigation

The following section of human factors information (pages 60-66) is taken from the *US Air Force Guide To Mishap Investigation* (AFP 127-1 Vol I, Chapter 10, "Aeromedical Investigation", May 1987)

The US Air Force source provides some additional insight on human factors issues and an investigative approach to the application of human factors. This information also provides an excellent summary of the main principles presented in the Hawkins text *Human Factors in Flight*.

The following diagram presents an overview of the areas of human factors which must be investigated for possible cause in a typical military aircraft mishap.

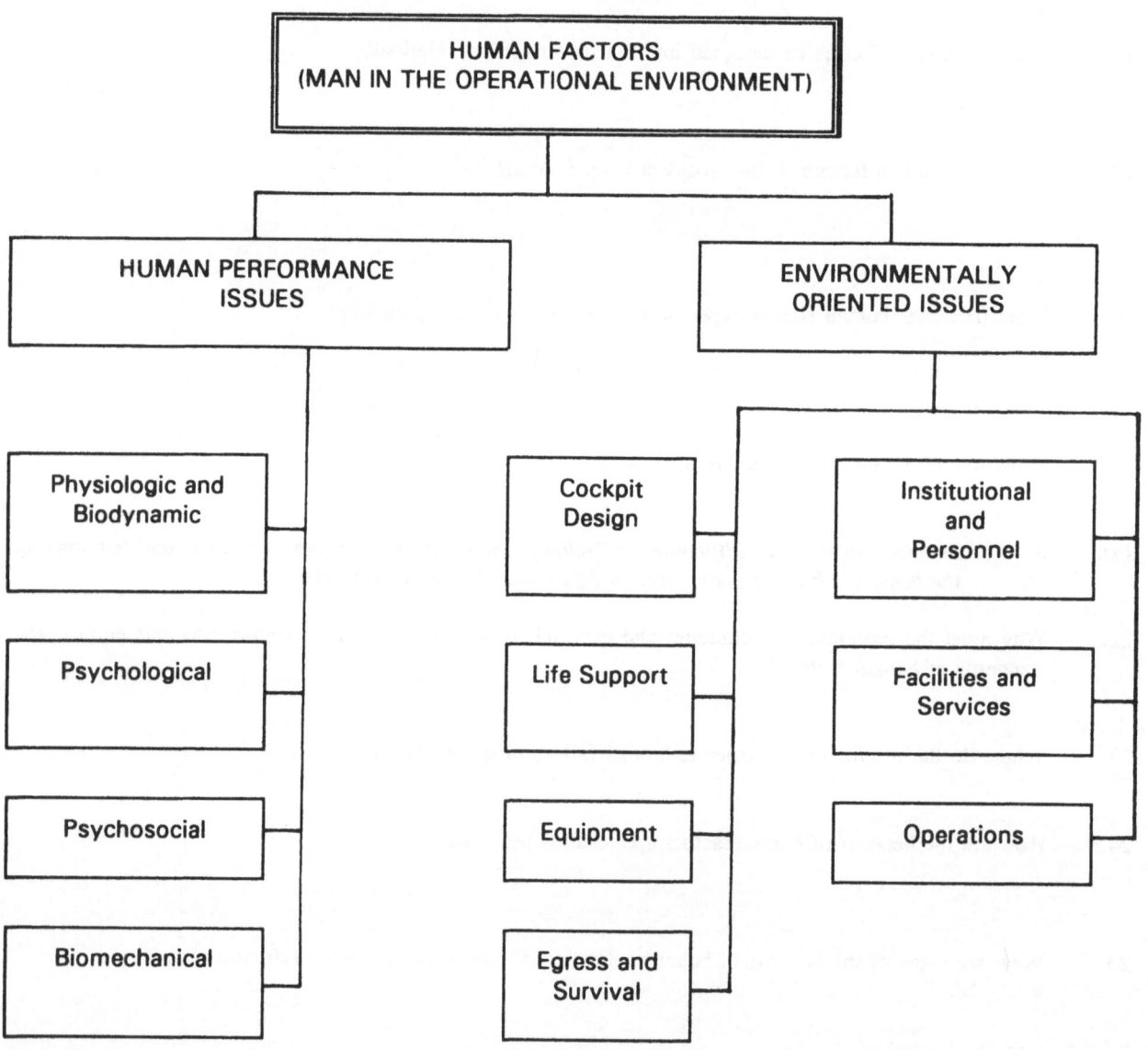

10-54. Human Performance Concerns. For purposes of both time management on the part of board members and completeness of topic and problem area coverage, the major areas of human factors interest are divided into man-centered human performance and environmental concerns. Human performance concerns are the traditional domain of the medical officer and his or her medical or psychological consultants, but must relate to environmental concerns addressed by other investigators. The issues considered during the medical data-consolidation phase should provide the basis for interaction with other board members who are integrating an overall picture of the mishap sequence during the team consultation phase. At this point, the need for specialized medical or psychological board augmentation may have been identified. As soon as a major issue in any human factors area becomes apparent, call HQ AFISC/SEL (AUTOVON 876-3458). The goal is to ensure a somewhat standardized and adequate analysis. This is not expected of the flight surgeon alone. The issues are often complex.

10-55. Physiologic and Biodynamic Concerns. These consider cardiorespiratory limitations, pressure change effects, the human senses, and various pathological conditions. A psychological training officer may be of some help in assessing these conditions. Acceleration effects on the cardiovascular system may include, among others, those related to the risk of inadequate blood flow to the brain, and those related to alveolar hypoxia under G. Hyperventilation is common, and because of the nature of the anti-G straining maneuver, may predispose to loss of consciousness along with its other effects. Pressure change effects should be considered to include hypoxia, evolved gas disorders, and various trapped gas effects. Various conditions of pathologic physiology as reflected in medical records, toxicology results, witness statements on health and fitness, thermal stress, and any possible causes for sudden incapacitation are basic and important issues.

10-56. Psychological Concerns. There must be continual efforts to improve our understanding of the factors which impact perception, information processing, and response. Behavior is not often self explanatory, and so it is essential that detailed information be gathered whether or not it seems pertinent initially. General problem areas include training, perception, attention, perceived stresses, fatigue, coping styles, and psychomotor capabilities. Although preliminary assessment of some of these

areas can be made by the flight surgeon, a team analysis of these areas is required. Only then can a meaningful integration of the different problem areas be achieved. When it is apparent that a psychologist would be useful to the board, there are some general criteria to consider in selection. An outline of these is provided in figure 10-22.

a. The concept of error pattern evaluation does not involve character judgement. For a multitude of reasons, simple human reality is that some behaviors are inappropriate to the task at hand. To intervene effectively, the specific associations must be teased out and studied. In essence, an error is seen from the standpoint of the competent investigator not as a cause so much as a manifestation of complex interaction between determinants of behavior. Technical errors such as missed radio calls or poor altitude or airspeed control are considered commonplace in aviation. Their reduction is the target of a variety of forms of proficiency training. Judgmental errors, on the other hand, involve more cognitive processing with the consequent choice consisting of a course of subtasks which may be inappropriate but only because they are a follow on to an inappropriate decision or judgment. These two types of errors are well known in safety, but are now undergoing further scrutiny by researchers in human performance.

(1) One type of distinction between types of human error is that between "slips" and "mistakes" (Morris and Rouse, 1985). Slips are characterized as errors of action. An example of this type of error is reaching for an automobile turn signal and finding the window wiper switch. Mistakes are discussed as errors of intention. For example, a driver stranded by a broken car with a history of electrical problems may replace the ignition computer rather than a broken fuel pump.

(2) Slips are thought to have several characteristics. They occur during some well rehearsed or established routine, appear to be associated with distraction or preoccupation, and flourish in familiar environments with few departures from the expected. Slips seem to be the result of actions that are highly automatic and so not consciously monitored. As a result, the expert may be even more prone to slips.

(a) The first and most common type of slip is where a habit pattern interference occurs. This is a response set where there is a change in routine (i.e., failing to stop at the store on the way home from work), a change in environment where the routine has not changed (i.e., walking to sit in a chair that has been removed), or the behavior is influenced by

61

environmental features (i.e., putting on one's coat instead of getting the box off the closet shelf).

(b) Unusual or ambiguous situations may facilitate a similar habit pattern phenomenon, a perceptual set. The perceptual set is the input side of this equation. A reading from an instrument may be that expected. Frequency is a factor when uncommon objects are misperceived as common ones. Incongruity, where an object does not "belong" in a setting may lead to a failure of perception, as can psychic need (i.e., a hungry person may perceive an ambiguous object as food).

(c) Omission or repetition of steps in intended sequences of events are "place losing errors." They are the result of systematization turned to ritualization (i.e., the use of a preflight checklist, missing, repeating, or not recalling a step).

(d) Slips can become distractions when they are recognized, and so perpetuate the problem. They can be expected to occur when environmental cues are not relevant to a present intention, when the environment has changed (cockpit configuration) but the task has not, when environmental features have not changed to facilitate an intended change in routine, when a long series of simple actions are required to complete a task, when the time between related actions is long or interrupted by other activity, or where procedures to do different tasks are similar in certain parts.

(3) Mistakes are more involved with judgment and decision making. Apparently well documented issues here are that decision makers consider no more than 2 or 3 variables at a time, and that recall may be triggered by prominent but irrelevant environmental cues. Attempts at solution may strive to incorporate the irrelevant information into the decision.

(a) The influence of past successes is probably disproportionately large and missing pieces of information are filled in based on individual theories which may have become reality to the decision maker. Once an operating hypotheses is selected, the natural tendency is to seek confirming evidence and explain away contradictions, often somewhat over-confident in one's state of knowledge.

(b) As a result of these characteristics of mistakes judgement, we can make some predictions. Conditions in which they are likely are: when more than two or three variables must be simultaneously considered, when prominent environmental cues suggest an inappropriate solution, when a particular inappropriate solution has been associated with success in past similar situations, and when choice of a solution requires approaching the problem in a novel fashion.

(4) Workload in this context is not viewed as a

cause of error, but rather as a catalyst to it. In other words, an increase in workload may not necessarily lead to error, but will be far more likely to be associated with one when other conditions are conducive to it. The increase in workload may be seen as a stimulus to improved performance on one hand and a distraction on the other. Whether one views a task as motivational or distressing will have considerable influence. Thus, issues of morale and motivation come into view. The occurrence of both mistakes and slips compound subjective mental workload. Both dealing with the consequences of an error (or correction) and knowing one has committed an error would weigh in.

(5) It is best to request consultant assistance when significant workload questions arise.

b. When considering aircrew training, a distinction must be made between the quality of the program providing it and the knowledge level of the recipient at some later point in time. Learning ability, rate, transfer, and practice or rehearsal are interrelated. An understanding of memory, including immediate (working), short term, and long term must be used in assessing a response set or habit pattern. A pilot's skill and knowledge must be compared to what is provided in terms of procedural guidance and training programs (be they ongoing local proficiency, upgrade, or initial) to draw meaningful conclusions. A training inadequacy as a categorical finding deals only with a program. Other findings may relate to the individual. Remember that one-time exposure to information does not equate to knowledge or training adequacy. Expert and novice differences and the impact of automaticity should be considered. It may be useful to compare any findings here with what may be expected of comparable aircrew (not necessarily even in the same squadron, but certainly peers in mission). Survey aids can be more reliable than random personal interviews, but should be used carefully. See the Safety Investigation Workbook (volume III) for a sample model questionnaire and contact HQ AFISC to request assistance. The pilot or investigating safety board member input will be vital here.

c. Perception here is intended to refer to reading of sensory information rather than sensation itself. Cognition is here the process of integrating various sensory and internal cues. Cognitive flexibility will facilitate insight and efficiency and may increase with training and experience. Training will generally serve to reduce the cognitive or conscious information processing time required to accomplish a given task. It may create a "mindset," in this case a perceptual set that "expects" a certain environmental cue. A response set (a response essentially out of habit) is a related concept but at the outflow end of the decision process. Confusion may result from a breakdown of effective cognitive processes and it may lead to serious misperceptions. Cognitive saturation is, on the other hand, the result of the capacity for cognition being exceeded by the number of available pertinent cues. It is useful in part to think of cognitive or attentional resources as a single capacity entity. Yet, it is clear that the cognitive resources must be interpreted in light of stage of process (early vs. late), modality of processing (auditory vs. visual, central visual vs. peripheral, proprioceptive, etc.), processing codes (spatial vs. verbal), and attitudinal and stressor effects. This is by no means a complete picture as knowledge in the area grows daily.

d. Attention involves the mental process of directing cognitive resources. There is a limit to attentional or cognitive resources which varies among individuals as well as in an individual based upon the day or the situation. A focus of attention will consume some of this which leaves a margin of attentional reserve. When there is no reserve, there is cognitive saturation. Distraction, whether from the consciousness or the environment, interferes with attention. Fascination is seen when attention is arrested during a crisis situation. The result may be "shock" or a "freeze" behavior. Temporal distortion may be associated with a high stress situation. During high stress, time may be perceived as moving much more rapidly or slowly than is actual. Awareness of time is distorted. Channelized attention occurs when attentional resources are focused on a limited number of environmental cues of subjectively high priority. Inattention is, on the other hand, the failure to focus attention appropriately. It is here that the difficult problem of repetitive task effects (such as boredom and complacency) comes into view. The general form of inattention is associated with boredom or complacency. The selective form of inattention, on the other hand, results from lack of knowledge or an inappropriate set of expectations. It is apparent from this brief review that more commonly discussed issues such as unrecognized or Type I spatial disorientation must be evaluated in conjunction with these topics.

e. Perceived stresses are influential but in an individual sense. Expectations an individual holds regarding his or her environment are important. Consequences one believes are contingent upon performance (written or not) will guide one's behavior. When actuality falls short of expectations or when positive expectations go unfilled over a long period, distress may be experienced. Emotion, insight, perceived expectations of peers, supervisors

or family, confidence in one's capability to deal with a situation, and perceived general workload all are powerful influences upon stress experienced as distress. The general adaptation syndrome (Selye) offers three phases in considering stress response. First is the fight or flight response which is immediate and temporary. Next, during the stage of resistance (the common stage), coping reserves are being actively directed toward adaptation in some form. Since this coping effort is active, it leaves diminished reserve. The final stage is that of exhaustion when the finite coping resources have been exceeded and symptoms my reappear.

f. Fatigue cannot be isolated from considerations of stress. Cumulative performance decrements can be a result of functioning in the stage of resistance over a long period. Acute performance decrements may be the result of high physiological and mental stress without adequate rest over some shorter period. Physical fatigue generally considers musculoskeletal limits of endurance. Sleep deprivation refers to an acute disruption of rest habits for whatever reason. Circadian cycles have a defiinate impact and must be considered based on "home" time. Motivational exhaustion refers to the emotional or affective component of fatigue and has a great deal to so with perceived stress experienced as distress. Biorhythms in the popularized good day/bad day form are not considered a proven entity. However, there may be a reason why stress is increased during weekend or holiday duty. Fatigue effects are pervasive, diminishing efficiency of mental processes from perception through exercise of judgment. Quantification can, however, be difficult.

g. Coping styles help to describe how an individual meets environmental demands. Decision making which includes the exercise of judgment and the selection of a response is further considered here. The actual mental process occurring is variable from one person to the next and from one time to the next. Studies have shown that what and individual describes about his or her decision is not necessarily accurate in reflecting what cognitive processing may have taken place. It may not be the same over time. Mental modeling is a complex issue. Inconsistencies may have nothing to do with integrity. This area will be a subject of continued research. It is important to keep this in mind in dealing with a witness or survivor interview.

(1) Personal discipline, general self confidence, motivation and other personality variables may play a role. Personality characteristics should be assessed but interpretation of this type of information must be

both candid and in consciousness of the biases of the observers. Such data may contribute to an explanation of why, for example, and individual over committed himself to a task.

(2) On the other hand, generalizations based on personality traits gathered in association with mishaps that might guide Air Force selection or other policies are unlikely.

h. Similarly, a subjective assessment of an individual's psychomotor capability may lead one to suspect a problem concerning strength or timing with control application. This, of course, must be done in light of what is "normal." Standards of normal may be broad and hard to apply even if available. Reaction time, for example, requires perception, diagnosis, exercise of judgment, selection of a response, and execution of the response. The time will vary depending on perceptual expectations, prior experience at exercising similar types of judgment, prior knowledge of alternative responses, practice experience (consequent skill) at response execution, and various factors that may diminish the proportion of cognitive resources that must be directed toward a given task. The time required to properly identify a situation and produce an unpracticed response may be as long as 6- to 9-seconds. This may have critical consequences.

10-57. Psychosocial Concerns. Here is where other supporting roles in the mishap may be exposed. Personal or community factors, supervisory influences, peer influences, and communication are among these concerns. The job environment is thought to be the more important area of focus. Because literature suggests this is a more proximate determinant of job behavior, the data may be more reliable, and some more direct intervention may be considered. (See paragraph 10-56 for comments on assistance.)

a. A person's perceived position within the community is again a matter of his or her expectations. These expectations of the environment in which the individual functions may not be easily assessed, but clues may be based on background such as education, travel, hobbies, religion, and career plans. Job satisfaction may be related to these, and to the extent to which the individual internalizes the values of the organization he or she purposes to serve. A powerful influence in this area may be close friends and family.

b. Supervisory issues are significant both psychosocially and institutionally. Command and control staff may have a powerful influence on behavior both by directives and enforcement of discipline, and by modeling (behavior that sets an example). When an individual has been directly tasked to meet a standard, the pressure is expressed. However, it is perhaps even more important to recognize the numerous and powerful influences exerted by supervisors by their behavior (verbally and physically).

c. Peer influences are even more heavily weighted toward learning vicariously (by observation). What happens to one aircrew will be closely scrutinized by bright and observant fellows. Verbal peer comments only partly in jest may constitute powerful influences on behavior.

d. Communication concerns include personal habits in communicating, intracockpit information exchange, information exchange beyond the cockpit, and communication equipment failure. Cockpit resource management is a term that describes the pilot as a manager of all his or her resources. This becomes far more significant in a multicrew aircraft where task delegation is accomplished by communication. As a result, personal habits in communicating (including message generation, intonation, and listening) become critical. What behavior is professional and effective in crew coordination should be addressed both from the viewpoint of the aircraft commander and others in the cockpit. For any aircrew, the quality of information gained by communication with outside agencies can be vital. This may also be a concern in an environment where cluttering and confusion on the airwaves interferes and may also impact interaction between flight members.

10-58. Ergonomic and Biomechanical Concerns. These are an area of traditionally intense effort on the part of the medical member, and have been discussed as a part of autopsy consideration. Based on team investigative progress, there may occasionally be a need to return to x-rays to find evidence of sabotage, to photographs to interpret man-cockpit contact, or to toxicology to confirm the influence of cockpit smoke or fumes. As a result, the need for careful early management of perishable evidence is reconfirmed. Body habitus, size, and strength should also be evaluated where it is practical to do so. MIL-STD-1472C provides body size measurement data (and depicts the points of measurement) by fifth to ninety-fifth percentile for men and women.

10-59. Environmental Concerns. The other safety board team members are each experts in their own right. As a result, exchange between each of them and the flight surgeon may cover these environmental concerns in light of the physiologic or biodynamic, psychological, phychosocial, and anthropometric or biomechanical information garnered. This section is an outline of the concerns each of those board members address as a part of their safety or human factors analysis. The Safety Investigation Workbook (volume III) includes a summary sheet designed as a convenience aid to recall factors discussed with other board members when the time comes to assemble an overall report. The general concerns for each key word factor are presence, contribution, degree of contribution, and temporal role.

a. General topical concern areas are broken down into specific problem concerns. Key words within a problem or subproblem area may have somewhat overlapping meanings. Again, an attempt has been made to place them in an 11 topic hierarchy in a fashion that will make it easier to understand the various relationships. The terminology may undergo some evolution, but a glossary is included in the Safety Investigation Workbook.

b. The process of board consultation outlined here will facilitate a comprehensive and coherent analysis of pertinent human factors. Other consultants may have been called in for special problems as well. Integrating these various perspectives and extracting conclusions can improve reliability. Credibility of recommendations for either immediate measures (as generated by the medical board member), or those as a result of data trending over a number of mishaps, depends on the quality of investigation by the medical member and his or her fellow board member consultants. As a result, current references and consultants should be used freely to supplement the outline provided by this guidance. Again, call HQ AFISC/SEL (AUTOVON 876-3458) whenever an area of human factors concern is identified.

10-60. Cockpit Design. This is a difficult challenge in small, multirole high-performance aircraft and in heavily automated multiplace cockpits. The problems may be weapons system specific and, as a result, the pilot member who is current in the mishap-type aircraft should review with the fight surgeon such problems as seat position, visibility, instrumentation, automatic systems and switch and control location. The idea is to assess the possibility for physical task saturation or to identify "designed in" impediments or limitations on mission accomplishment. (Accurate information is very important to designers.) Examples of problems in this area are numerous. The heads up display has been seen as a new problem area. The Instrument Flight Center at Randolph AFB TX is commissioned to study some of these problems, Aeronautical Science Division at Wright Patterson AFB OH Among others, also researches these issues.

10-61. Operations Concerns. Flight-specific concerns should also be covered with the pilot member (and perhaps others). Mission demands begin with planning and briefing. Special flight stresses, such as range operations, aerobatic confidence maneuvers, various air combat tactics, and the possibility of acceleration displacement effects, should be more pertinent to fighter types. Special navigation, NAVAID, fatigue, or automation problems may well be more pertinent to larger aircraft. Also to be addressed are weather or night problems, emergencies (and pertinent emergency procedures), potential toxic exposures unique to an aircraft, and potential incockpit trauma.

10-62. Life Support and Personal Equipment. Adequacy of this equipment to meet the special demands and risks of flight have been much improved through the mishap analysis of life-support officers and flight surgeons. This is a historic model for consultation to the flight surgeon. Both individuals discuss the cockpit environmental control and oxygen delivery systems, anti-G or pressure suit equipment, helmets, special mission gear (such as CBW), and other items of personal clothing and equipment. The life-support officer, when present, is normally a nonvoting board member.

10-63. Facilities and Services. They should be discussed to include quality, availability, and any relationship to the mishap. The investigating safety officer will be a consultant in this area. The use of facilities is more a psychological or psychosocial concern, and access should also be considered here. Access to adequate nutrition, quarters for rest, and exercise, recreation, and health care facilities should be reviewed for any potential influence on a mishap sequence. In a more direct way, the facilities of an airfield or base, such as field lighting, weather service, aircrew dispatch, special intelligence, rescue or fire control services, or air traffic control, may also play a role.

10-64. Equipment Concerns. Consider at the level of both local maintenance and that of the logistics system. The flight surgeon will assist the maintenance officer in evaluating local human concerns affecting maintenance. Many of the physiological, psychological, psychosocial, and even some of the anthropometric concerns enumerated may apply to the maintenance specialist and his or her supervisors. The important basic categories in which these principles may apply are evaluation of field quality assurance and field working conditions (including tools and facilities). In addition, unit manning and individual qualification may be important. The maintenance officer may have additional questions concerning logistical considerations he or she must address, such as depot quality assurance, depot management, acquisition or modification philosophy, overhaul philosophy, or design defiency. However, these latter topics are the domains of systems safety.

10-65. Institutional or Management Issues. This concerns policies that may be shown to have had some relationship to a mishap. The experience and expertise of the board president make him or her a valuable consultant with whom to review issues of selection, evaluation, promotion, additional duties, the military or locally unique lifestyle, and internalization of unit and organizational values. Other board members may also be consulted, but it is appropriate to discard random speculation. Professional military education exposure and perhaps carefully configured surveys may be of some use.

10-66. Egress and Survival. Address all of the various man-centered concerns in the time frame after the point of the mishap. This point is the time when the mishap becomes inevitable regardless of crewmember action. In an ejection seat aircraft, there may be a need to assess the timeliness of an escape decision. Once again, the life-support officer is the consultant.

a. Without an ejection seat, the point of the mishap may not be so readily established. (It is not always the time of impact.) However, there is often still a phase where aircrew efforts are directed to either escape or survival. Human factors concerns in this phase are not to be confused with those which play some role leading up to the point of the mishap. Analysis of one group of human factors may lead to mishap prevention, while the other may be studied to improve survivability of the mishap once it occurs. Injury pattern analysis is often used to assess the effectiveness of egress systems. Often, the clue is as small as a paint smear on a piece of clothing. The life-support officer is a key board member in this analysis. Volume II addresses techniques of value in assessing aircraft crashworthiness for comparison with human impact tolerances. In many high-speed mishaps this is not required.

b. Once an escape has been accomplished, the many potential problems of survival, such as water, heat, cold, or first aid, may come into play. The final act is that of rescue, and its effectiveness must also be assessed. It would be best to note who accomplished the rescue, whether and when physician supervision was employed, whether rescue personnel were deployed and adequately trained to meet the demands, and whether survival care was adequate.

Human Factors Investigation

PO: *Describe and give examples of the "Human Performance" and "Environmentally Oriented" issues of human factors investigation as described in US Air Force Guide To Mishap Investigation (AFP 127-1 Vol I, Chapter 10, "Aeromedical Investigation", May 1987).*

1.   With what does human factors deal?

2.   What are the two main issues to consider when investigating human factors?

3.   What are the "Human Performance" issues of US Air Force human factors?

4.   What are some of examples of the physiologic and biodynamic human factors concerns?

5.   What are some examples of the psychological human factors general problem areas?

6.   How is an error viewed from the human factors investigation standpoint?

7.   What is the difference between technical errors and judgmental errors?

- Technical   =

- Judgment   =

8.   What is the difference between a slip and a mistake?

- Slip   =

- Mistake   =

9.   What is cognition?

10.   What is attention?

11.   What is fascination?

12.   What are the psychosocial concerns the human factors investigator must examine?

13.   What human factors in the ergonomic and biomechanical areas must the safety investigators be concerned with?

14.   What are the "Environmentally Oriented" issues of US Air Force human factors investigation?

15.   What human factors of cockpit design must be examined in the safety investigation?

16.   What are the operations concerns of human factors the safety investigator must check on?

17.   What life support and personal equipment aspects of human factors does the safety investigator need to examine?

18.   Under facilities and services, what human factors must the safety investigator look into?

19.   What equipment concerns of human factors must be evaluated by the safety investigator?

20.   What are the institutional, training, or management issues that can have an impact on human factors?

21.   What are the egress and survival concerns of the human factors safety investigator?

Video Involvement Questions # 1

"ALERT 3" (Crash of United Flight 232)

This is a video recording of a presentation made by Capt Al Haynes to a group at the NASA Ames Research Center, Dryden Flight Research Center at Edwards AFB, CA on 24 May 1991. He discusses the crash of United Flight 232 at Sioux City, Iowa on the 19th of July 1989.

The introduction is a sound track of some of the ATC communications with United 232, then Capt Haynes makes his presentation.

PO   1.   *Answer the questions designed to highlight various details of the video presentation.*

PO: 2.   *Identify the aspects of human factors that are found in the presentation given by Capt Haines on the crash of United 232. Use the SHEL conceptual model to describe good and poor interface examples.*

1.   How was the crew controlling the turns in their DC-10?

     a. ailerons only   b. rudder only   c. power only   d. a combination of a & c

2.   Which direction of turns did the crew tell the controller they could make?

     a. Right   b. Left   c. Left or right

3.   Initially did the crew think they would make it to the airport?

     a. Yes   b. No   c. There were serious doubts

4.   What was the final result of the minor misunderstanding about elevator authority?

     a. They did have elevator control
     b. They had minimal elevator control
     c. They had no elevator control
     d. they had intermittent elevator control

5.   What directions did the crew give to the ground emergency response vehicles?

     a. Place them at the approach end of the runway.
     b. Place them at the departure end of the runway.
     c. Locate them at mid field
     d. Put them in the corn field at the end of the runway.

6.   How far were they from the airport when they put the landing gear down?

     a. At 30 mi   b. Outside 10 mi   c. Between 5-10 mi.   d. Inside 5 mi

7.   What runway did tower clear United 232 to land on?

     a. RWY 22   b. RWY 31   c. RWY 13   d. Any runway

Capt Haynes' Presentation

8.   What were the five factors Capt Haynes said made it possible for them to have as many survivors as they did?

     (1) _____

     (2) _____

     (3) _____

     (4) _____

     (5) _____

Factor # 1

9.   What did he say about getting the airplane on the ground with only two engines for control?

_____

10.   How did he feel about the weather and turbulence at the time of his emergency?

_____

11.   What part did the time-of-day play in the emergency?

_____

12.   How did the Air National Guard fit into factor # 1?

_____

Factor # 2

13.   What did Capt Haynes say about the controller who gave them directions to the field?

_____

14.   What did he say about radio frequency congestion after they declared an emergency?

_____

15.   How did Capt Haynes feel about the cockpit and cabin crew communications during the emergency?

_____

16.   What results did factor # 2 have on the emergency response forces?

_____

Factor # 3

17.   How did factor # 3 effect the response of the crash/rescue and other participants in the disaster?

_____

18.   What was the key to the success of the response of the cabin crew?

_____

19.   What cockpit crew training did Capt Haynes say was the most valuable for them?

       a. Simulator training in total hydraulic failure situations.
       b. Cockpit Resource Management (CRM)
       c. Line Orientated Flight Training (LOFT)

Factor # 4

20.   What did the crew discover would minimize the pitch oscillations?

_____

21.   What response error did Capt Haynes say the crew was making in the attempt to keep the wings level?

_____

22. Who ended up working the throttles?

    a. Capt Haynes   b. The 1st officer   c. The Flight Engineer   d. An extra DC-10 Capt

Factor # 5

23. How much combined experience operating the DC-10 did the cooperation of each of the cockpit crew total?

    a. 65 years   b. 90 years   c. 103 years   d. 115 years

24. What insightful way were the passengers prepared to increase the chance of survival for the children?

    _____

25. How many people from the surrounding areas did Capt Haynes say lined up to offer blood ?

    _____

Video Involvement Questions # 2

"WHY PLANES CRASH"

This video is a NOVA presentation, written, produced, and directed by Veronica L. Young. Executive producer - Paula S. Aspell in 1987. Coronet Films Video, 108 Wilmont Rd., Deerfield, IL 60015.

PO   1.   *Answer the questions designed to highlight various details of the video presentation.*

PO: 2.   *Describe the aspects of human factors that are found in the video presentation "Why Planes Crash." Use the SHEL conceptual model to describe good and poor interface examples.*

1. How often do airline passengers think of the pilot or crew being at fault in their natural concern for their safety?

    a. Frequently   b. About half the time   c. Rarely

2. How many people fly on planes in a year?

    a. 100 million   b. 500 million   c. 700 million   d. 900 million

3. How many people are killed world wide in plane accidents in a year?

    a. 800   b. 1000   c. 1500   d. 2800

4. What are the three things that are happening which pose a safety challenge for aviation?

    1. recovery from the _____

    2. _____ - new pilots and controllers in the system

    3. deregulation environment - more _____

5. What does Capt Roger Brooks say about why the margins of safety are reduced?

    - _____ has changed based upon increased _____ pressures

6.   What did John Nance say lead to the Air Florida Flight 90 accident in 1982?

      - _____

7.   What was the first human failure noted that caused the Air Florida 737 to not have enough power to continue the climb after liftoff?

      - The crew did not _____

8.   What other deicing procedures were also ignored?

      - _____

9.   What is the range of accidents blamed on cockpit error?

      - _____

10.   What responsibility did the first officer have in the Air Florida crash?

      - He knew something was wrong but _____ himself.

11.   What side of the "safety equation" did Capt Mel Volz say needed attention now?

      - the _____ side

12.   Where are the roots of some of the human factors problems on the flight decks of today?

      - in the macho _____ of the single seat fighter pilot

13.   What were some of the motivational feelings given for being a pilot?

      - feeling of exhilaration, _____ and _____

14.   What does Dr Clay Foushee, the aviation psychologist, say might be parts of todays' "right stuff" in the cockpit?

      a. concern with the _____ (technical skills)

      b. ability to _____ well to other _____

15.   What are the two aims of Cockpit Resourse Management?

      a. modify _____ to a degree

      b. get the crew to _____

16.   What does Dr Clay Foushee see as one of the most useful techniques in CRM training?

      - the full mission simulation with _____ of actions

17.   In CRM deficiency evaluation what, in addition to how well the crew handled the technical challenges, is also evaluated?

      a. How well the emergency ended   c. How precisely they controlled the aircraft
      b. Their coordinated effort   d. How responsive the ATC system was

18. What is one of the values coming from the "video feedback" in CRM?

- allows one to see how _____ as a crew member

19. What recurring problems are illustrated by the disaster of Eastern Flight 401 which crashed int he Florida Everglades?

- failure of captains to act as leaders, _____,

set _____, and _____ responsibilities

- shortcomings which are compounded by _____ and _____ crewmembers

20. What did the captain neglect to do which allowed everyone else on the crew to be absorbed in the crisis?

a. Declare an emergency
b. Communicate the nature of the problem
c. Divide up flying responsibilities
d. All of the above

21. What was wrong with the controllers inquiry "Eastern 401, How are things coming along out there"?

a. It was too vague
b. It didn't communicate to the crew the real concern
c. It didn't alert the crew to the altitude deviation
d. All of the above

22. What was one reason given by a first officer for not making mention to the captain of an observed problem?

a. The captain could not understand the language.
b. There was too much noise in the cockpit.
c. The captain would really "jump" on him.
d. The captain would not listen anyway.

23. What does Dr John Lauber say his point of view is about CRM training?

a. It is a refined method of training that is proven to be very effective.
b. It is a concept still in development.
c. The training concept has received universal acceptance in the airlines.
d. Both a and c were stated.

24. Which of the airlines (at the time of this video) had full fledged CRM training?

a. Pan Am   b. United   c. Delta   d. Southwest
e. Eastern   f. Peoples Express   g. Continental Express

25. According to Dr Clay Foushee why are accidents a terrible research criterion for evaluating success of any program?

a. because they are so infrequent
b. because there are so many fatalities with each one
c. because they are so frequent
d. because human factors are so hard to determine

26. Why does Del Fadden say the pilot will not be eliminated from the flight deck in the foreseeable future?

- The pilots skill at _____ _____ _____ can not be duplicated

27. How does the Boeing 757 and 767 relieve the pilots' mental workload?

- by using a computer which _____ decision making

28. How does the computer system of the glass cockpit save mental effort of the pilot in decision making?

- by _____ the _____ needed for a _____

29. What does Dr Earl Wiener say are the two "D" problems of any automatic device?

a. Delicate and dangerous   c. Dutiful and delicate
b. Dumb and dangerous   d. Dumb and dutiful

30. What is one of the human factors problems of monitoring automation?

- _____ or lack of vigilance

31. What does Dr Wiener say is a very realistic concern the pilots have about the increasing use of automation in the cockpit?

a. They will loose their jobs
b. The will have a loss of skills
c. The computer will make an unrecoverable error
d. Computers just can't fly as well as the human pilots

32. What does John Nance say the problems in the aviation industry are based on?

- lack of _____ _____ of how close you can come

to the line and not _____ _____

Video Involvement Questions # 3

"THE WRONG STUFF"

This video is a HORIZON presentation, written and produced by Jeremy Taylor. Film editor - Peter Essex. HORIZON editor Robin Brightwell

PO   1.   *Answer the questions designed to highlight various details of the video presentation.*

PO: 2.   *Point out the aspects of human factors that are found in the video presentation "The Wrong Stuff."
Use the SHEL conceptual model to describe good and poor interface examples.*

1. What does the video "The Wrong Stuff" focus on?

a. Flight crew behavior
b. Failures of the ATC system
c. Problems of new technology
d. Modern weather problems

2. What are the two very important facts about air safety that come from examining crashes?

   1. Commercial aviation is still incredibly _____.

   2. Accidents _____ happen and in about _____ of them the pilot is to blame.

3. In the famous incident of the 747 at Nairobi cited by Roger Green, why did the crew choose not to believe the warnings about being below the glide slope?

   a. Visual cues verified the warnings were false
   b. There was disagreement among the crew
   c. The crew was distracted by an engine failure
   d. They did not fit into their model

4. According to Dr John Lauber's review of the 707 crash into the mountain on Barla Indonesia while attempting an NDB approach, what did the crew fail to do?

   a. recognize station passage
   b. set in the correct altimeter setting
   c. positively resolve the ambiguities
   d. verify an obviously wrong assigned altitude by ATC

5. In the Boeing 727 simulation with a No 3 engine fire, what error in the cockpit resource management did the captain make relative to the flight engineer?

   a. gave him the wrong configuration
   b. overloaded him with tasks
   c. did not accept his vital inputs
   d. sent him back in the cabin to fight the fire

6. According to Dr Bob Helmreich which of the 3 components at the test pilot right stuff creates problems when trying to function as an effective team?

   a. High technical competence
   b. Rugged individualism
   c. High level of competitiveness
   d. Both b and c above

7. What seems to be the ironic result from the lack of copilot assertiveness illustrated by accidents like the Air Florida crash?

   a. they would rather die than stick their necks out
   b. they become over assertive captains
   c. they can't read engine instruments as well as captains
   d. their flying skills are better than most captains

8. When captains in a simulator study pretended to be incapacitated, what percent of the simulators crashed when the copilots failed to take over?

   a. 10%   b. 15%   c. 20%   d. 25%

9. As a result of the United 707 crash at Portland how did the airline rethink its approach to pilot training?

   a. It is basically a _____ _____ problem.

10. In the United CRM training discussion how is the 9-9 management style defined?

    a. advocates his position
    b. inquires from the rest of the crew
    c. concerned with accomplishing the task
    d. concerned with input of the crew
    e. follows through on conflict resolution
    f. can make a decision
    g. all of the above

11. In the review of the video replay of the simulator flight for the United CRM training, what did the instructor point out was a significant thing being done?

    a. the captain was dominating the situation
    b. the crew was focusing too much on the emergency
    c. there was a good division of responsibilities
    d. the copilot was not as assertive as he should have been

12. What did Peoples Express focus on to eliminate the wrong stuff from their cockpits?

    a. extensive use of cockpit resource management
    b. selection screening for people who can work with people
    c. line of flight training (LOFT) for flight and cabin crew
    d. selection of good looking people with good technical skills

13. What has drastically changed the nature of the pilot's job in modern transport aircraft?

    a. computer automation
    b. higher performance engines
    c. wide bodied aircraft
    d. using only 2 pilot crews

14. What does the "Electronic Cocoon" do to help strike the balance between human and machine?

    a. allows the crew more freedom to operate
    b. provides warning to the crew when they approach limits
    c. creates a dialogue between the human and machine
    d. all of the above

15. Why does Dr John Lauber say the computer will never replace the pilot on the flight deck?

    a. only people can be creative
    b. people can fly as good as computers
    c. people will never trust computers totally
    d. it's better to have a captain than a computer first on the accident   scene

Video Involvement Questions # 4

"TOP GUN AND BEYOND"

This video is a NOVA presentation, written, produced and directed by Chris Haws. Executive Editor - William Grant. Executive producer - Paula S. Aspell

PO   1.   *Answer the questions designed to highlight various details of the video presentation.*

PO: 2.   *Discuss the points of human factors that are found in the video presentation "Top Gun and Beyond." Use the SHEL conceptual model to describe good and poor interface examples.*

1.   What are the engineers of today's fighter aircraft not adequately dealing   with?

      a. Psychology   b. Technology   c. Biology   d. Statistics

2.   At the speeds of today's modern fighter aircraft how much time is available for a fighter pilot to deal with the enemy who is 20 miles away?

      a. 15 sec   b. 30 sec   c. 1 min   d. 2 min

Next there will be about a 15 minute historical development of fighter aircraft technology.

Make note of any human factors considerations indicated during this period.

3.   What was the result of all the various inputs the pilots of the Vietnam era had to deal with?

      a. Pilot saturation and overload
      b. Cockpit display design improvements
      c. Peak pilot psychological performance
      d. The best designed aircraft ever developed

4.   What happens to a fighter pilot if he does not unload and get the blood back in his head after experiencing tunnel vision and first blackout?

      a. He sees black and white dots.
      b. He experiences "red out".
      c. He experiences a loss of consciousness.
      d. He has heart failure.

5.   What is the estimate of pilot deaths in 5 years from "G" loss of consciousness?

      a. 5   b. 10   c. 15   d. 20

6.   How long does it take for the higher learning, cognitive centers of the brain to start working adequately after a "G" loss of consciousness incident?

      a. 12 sec   b. 30 sec   c. 2 min   d. 10 min

07.   How many extra "Gs" of tolerance can the anti "G" straining maneuver bring the pilot?

      a. 1   b. 3   c. 5   d. 8

08. What new study is being conducted to help the aircraft monitor the status of the pilot?

   a. Brain wave monitoring
   b. Blood pressure monitoring
   c. Eye movement monitoring
   d. Breathing monitoring

09. What are the "smart systems" being developed at Wright-Patterson AFB designed to help the pilot with?

   a. elevated brain blood pressure
   b. better instrument scan patterns
   c. greater situational awareness
   d. improved aircraft control inputs

10. With all the technological developments, why can't the computer replace the pilot?

   a. the human flexibility advantage
   b. the computer visual deficiency
   c. the computer lack of processing speed
   d. the lack of computer accuracy